■

Super Blue

Photography by Katsuhiko Tokunaga

■

for Lieutenant Yves "Kikou" BRO/Pilote, Patrouille de France (1985~1988)

■

HOWELL PRESS

At 30,000 feet altitude, the air may be as cold as -50°C. The sky is perfectly clear, an endless blue. Sleek aircraft, like enchanted sonic birds in a dark blue sanctuary, streak through the extreme cold on metallic wings.

■

德永克彦
（とくながかつひこ）
Katsuhiko Tokunaga

1957年1月13日生れ。東京都出身。1978年9月のT-33A同乗
以来、各国軍用機の空対空撮影を中心に、積極的に取材活動
を続けている。主な発表媒体は内外の航空専門誌。他に海外
航空機メーカー、空・海軍等の公式写真の撮影でも実績を積
み、近年は、航空ビデオの製作・監修も多く手掛けている。

Katsuhiko Tokunaga was born January 13, 1957. Ever since he first rode in a T-33A jet trainer in September 1978, he has covered the military high-performance jets of many nations, concentrating on air-to-air photography. His published work has appeared in magazines in Japan and overseas. In addition, he has taken official pictures for foreign aircraft manufacturers, air forces, and navies. He has also worked on the production and direction of many aviation videos in recent years.

■

SŌKŪ NO SHIKAKU/SUPER BLUE by Katsuhiko Tokunaga.
Copyright © 1987 by Katsuhiko Tokunaga.
English translation rights arranged with CBS/Sony Publishing, Inc., through Japan Foreign-Rights Centre.

English language edition published in the United States by Howell Press, In 2000 Holiday Drive, Charlottesville, Virginia 22901. Telephone (804)977-4006. All rights reserved.

This book, or any portions thereof, may not be reproduced or transmitted in any form or by any means, electronic or mechanical, including photocopying, recording, or by any information storage and retrieval system, without permission in writing from the publisher.

Printed in Japan.
Library of Congress Catalog Card Number 88-80879
ISBN 0-943231-15-9

■

HOWELL PRESS

CBS・ソニー出版

蒼空の

Super Blue

視覚

撮影・德永克彦

Photography by Katsuhiko Tokunaga

CBS・ソニー出版

I admit that this is rather personal, but 1987 was a commemorative year for me—it marked 10 years since I started my career as an aerophotographer.

Because I began taking photographs as a hobby, and my hobby eventually developed into my profession, it is hard to pinpoint the beginning of my career. I believe the summer of 1977 was my starting point: at that time I had an enjoyable experience doing aerial photography in the United States for *Koku Journal*. Although the past decade has not been an easy path, thanks to the help and understanding of a number of individuals, especially the editorial staffs of many aerospace journals, I was able to develop my career. I would like to express my gratitude to these people.

Someone might get the impression that I spent the past decade without any purpose, but that is far from true. On the contrary, it is not an exaggeration to say that I have devoted my life to aerophotography. Compared with the photography of automobiles and railways and so forth, the photography of aircraft is a minor area. Aerophotographers are generally called "pseudophotographers" in avia-

tion terminology, because they photograph the ground from an airplane.

You can imagine what it was like 10 years ago—conventional wisdom said that aircraft could be the means for taking pictures, but could not be the subject of pictures. This thinking resulted from the fact that aircraft are not as popular in Japan as they are in the U.S. Until recently, there were few cameramen who specialized in aerial photography; therefore, it was rare to find quality photographs. While the situation was regrettable, it worked to my advantage when I began work as a cameraman. I did not earn much at first, but I have been able to support myself from the very beginning.

Most of the subjects on which I worked were unexplored. I felt a measure of satisfaction each time I finished a shooting, and that was a big incentive to continue. Although the airplane was a total stranger to me when I began doing aerophotography as a hobby, it became very familiar when I focused on it as a subject for my profession.

In the beginning, the enjoyment of working in a new field that I had not been able to pursue before was

greater than the satisfaction of obtaining a good picture. It is natural for a professional to claim that his first priority is to achieve good results, but, to tell the truth, it is hard to say if I had that awareness in my first few years as a photographer. Rather, things like being exposed to a new machine and going to a new place were my main interests. Even now, I do not deny that I am an "airplane lover," and that I find working with airplanes thrilling. I get excited, of course, about the new countries and new places, and I enjoy, even now, my special feeling toward new aircraft. I realize that I am becoming more self-conscious as a professional now, especially when I see myself thinking about how I should compose an image.

It was the aerobatic teams of many countries who gave me the greatest excitement for several years, and you'll see them in this book. I have taken the photographs of eight teams so far, starting with the U.S. Air Force Thunderbirds in 1980. Each team had a distinctive personality and was a totally absorbing experience for me. I could not list all the fascinating characteristics of the aerobatic teams here, but if I were asked to say what is most impressive about them, I would have

to point to their precision-flying ability.

Military aircraft and their equipment are beautiful, but we cannot ignore the fact that they are weapons. Being a non-political photographer who was simply attracted by jet fighters, I do not seek either to make an anti-war statement or to support armaments. My blood runs cold, however, when I visit a country and the people say, "We are for our country, right or wrong." That strong nationalism concerns me because I am a member of the post-war generation.

There are exceptions, but almost every team shows its greatest interest in "flying." Of course, there is no doubt about the fact that the main purpose of each team is to recruit others to join its armed forces. I feel most comfortable working with the teams who concentrate on improving their aerobatic technique, while freely admitting that their main purpose as an organization is to recruit others.

It is natural that I should want to photograph loops, rolls, and big formations when I ride in a powerful jet fighter. No matter how powerful that aircraft is, however, there is very little difference between photographs taken from a high-performance

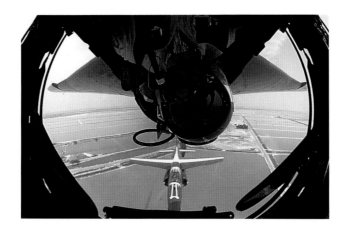

aircraft and those taken from a passenger aircraft, if yours is the only airplane flying. The tight formations of the aerobatic teams provided me with the opportunity to take special pictures. If you ever experience the thrill of pulling Gs in a 20- to 30-minute aerobatic show sequence, you will be a camera fiend.

There are two methods of photographing aerobatics: riding in one of the aircraft in formation, or riding in another aircraft that chases the formation. I prefer the latter. It is usually more difficult to obtain permission to do the former because it is more dangerous for the team. A passenger riding with an aerobatic team flying in close formation could cause a serious accident if he or she accidentally hits a control even for a second. If the individual is merely a passenger with no equipment, the risk of an accident is reduced. But if the passenger is an inexperienced photographer, carrying a camera which could become a lethal weapon during the flight, the team must be deadly serious.

The Red Arrows prohibit cameramen from riding in the formation. The same rule applies to the Frecce Tricolori, except for the aircraft that fly on the centerline of the formation. The Blue Angels and the Thunderbirds, the American teams, allow people to fly in the back seat of the spare aircraft if it flies alone, but they do not permit cameras, tape recorders, or other items in the cockpit. It is a big honor for a cameraman to get permission to ride with a formation; it means the team trusts you.

Once you get in the cockpit, the show sequence is already determined, and the picture opportunities develop by themselves. I simply sit in my ejection seat and wait for the climax, concentrating only on lens selection. It is a wonderful and invaluable experience to be able to see exactly what the pilot sees from the cockpit, but it is somewhat unsatisfactory for creating dramatic pictures.

The situation is very exciting, however, if you set up the flight sequence for yourself by photographing from an aircraft that chases the aerobatic formation. It is especially interesting to discuss the creation of each flight pattern with the pilots of the formation and the chase aircraft. If you were in the aerobatic formation itself, you would not have the advantage of pulling up to the best possible position to photo-

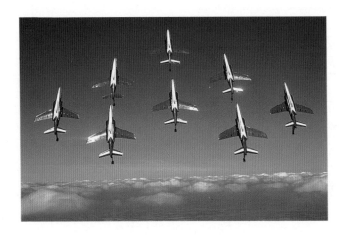

graph all of the aircraft in the aerobatic formation.

Usually an aerobatic show alternates between a formation and a solo demonstration in front of the audience. Since a jet's turning radius is so large, the alternating demonstrations provide two different shows to fill the time. Each demonstration covers up the space required for the turn of the preceding flight. Since I do not have many chances to fly, I want to take as many pictures as possible. That means right after I shoot a sequence on the formation, I will go after the solo performance, then back to the formation. I need to take a shortcut to the inside of the performing aircraft, which means I experience the heavy G-forces of a tight turn, and could lose awareness. The pilot also needs to make rapid accelerations and decelerations to maintain close proximity to the other aircraft.

Under these circumstances, the cameraman needs to take accurate aim without a moment's hesitation, and he must determine the position of the formation and the solo flights for the pilot, who is concentrating on chasing the next sequence of the performance. And, of course, he has to make sure he releases the shutter when the aircraft reaches its best position.

Both pilot and cameraman must have the highest technical skill and expertise. Nothing can take the place of the professional camaraderie between the pilot and the photographer during these flights. As a matter of fact, the trust between the pilot and the cameraman is the most important element of a successful aerophotography flight. When the relationship between the two is weak, the aerobatic team will emphasize safety to a greater degree. First, they drop the solo chase and allow only formation photography. Next, they drop all performances with high-G loads. The most drastic step is the prohibition of passengers on flights altogether.

Fortunately, this has never happened to me, but I have seen many cameramen refused this way. There is no doubt about the fact that aerobatic flight requires great concentration. If the pilot has to give even 10 percent of his attention to his passenger, the flight may no longer be safe. It might seem to be difficult for me as a foreigner to establish a relationship of trust, but it is not all that difficult. In a sense, it is much more challenging to establish a relationship of

trust with the teams, whose personalities differ from country to country, than it is to photograph their different aircraft.

In the summer of 1987, I spent three weeks with La Patrouille de France. When the work was done, I was extremely happy about the fact that I had built a strong working relationship with the team. The happiness that I felt because I had done a good job with the photography was secondary. The French Air Force team is among the world's leading aerobatic teams, one with which I had never before worked. In the tenth year of my career, it served as my commemorative work. Returning home, I was fortunate enough to be able to publish my work in this book, *Super Blue.*

There have been, of course, several aerophotog-raphy books published in the past. In Japan, however, they were just aircraft picture-books, rather than true aerophotography. This has always frustrated me as a photographer. And so this book of true aerophotography, published in the tenth year of my career, has claimed a special place in my heart. I would be pleased if through these pictures you are able to feel the functional beauty, intensity, and excitement of these airplanes.

The last and most important word of gratitude goes to Mr. Sohichiro Sano, the editorial director of CBS/Sony Publishing, who produced the Japanese edition; Mr. Hidetaka Mochizuki, of Hot Art, who did a wonderful job as art director; and Howell Press, publishers of the English language edition.

October 30, 1987   *Katsuhiko Tokunaga*

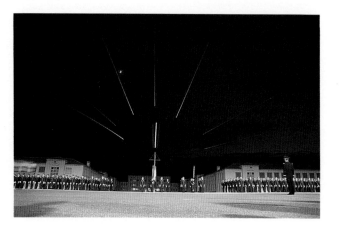

in the States

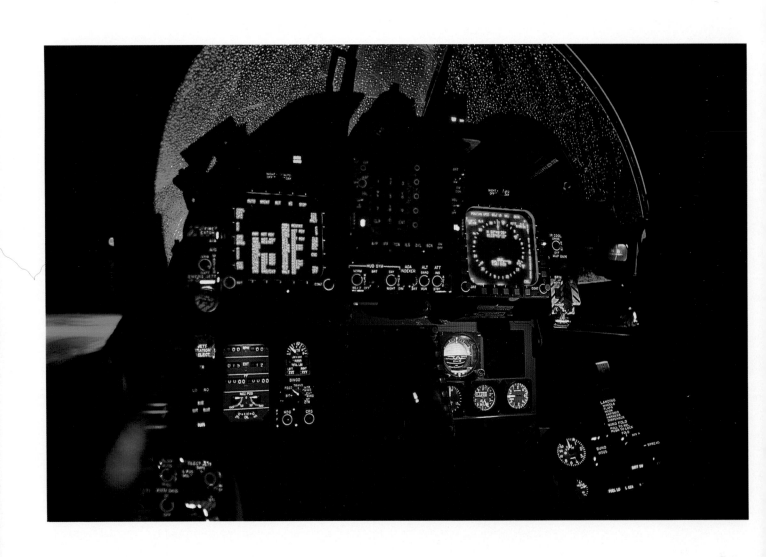

in Europe

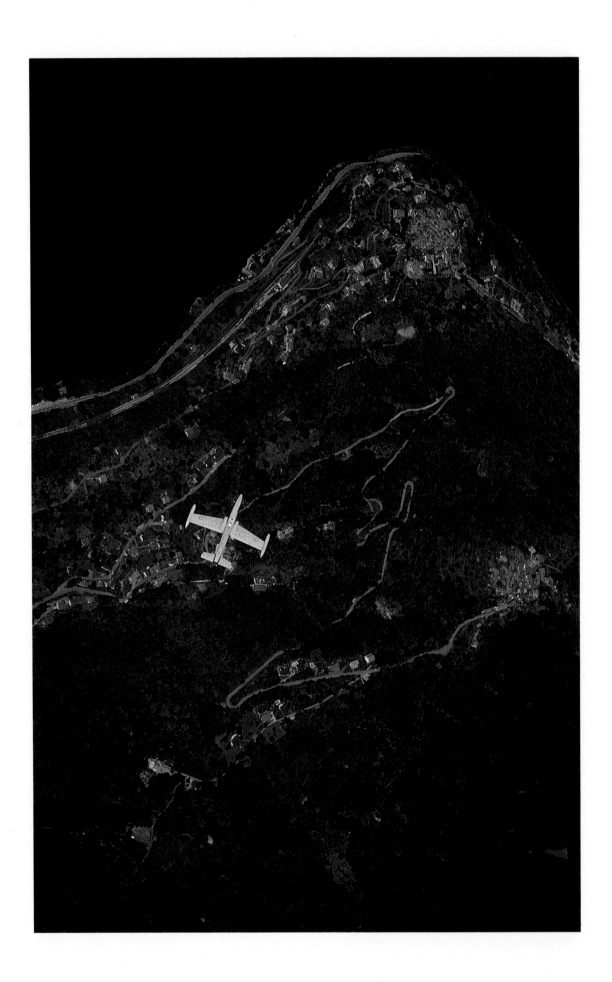

■

in the Far East

■

■

and Aerobatic teams

■

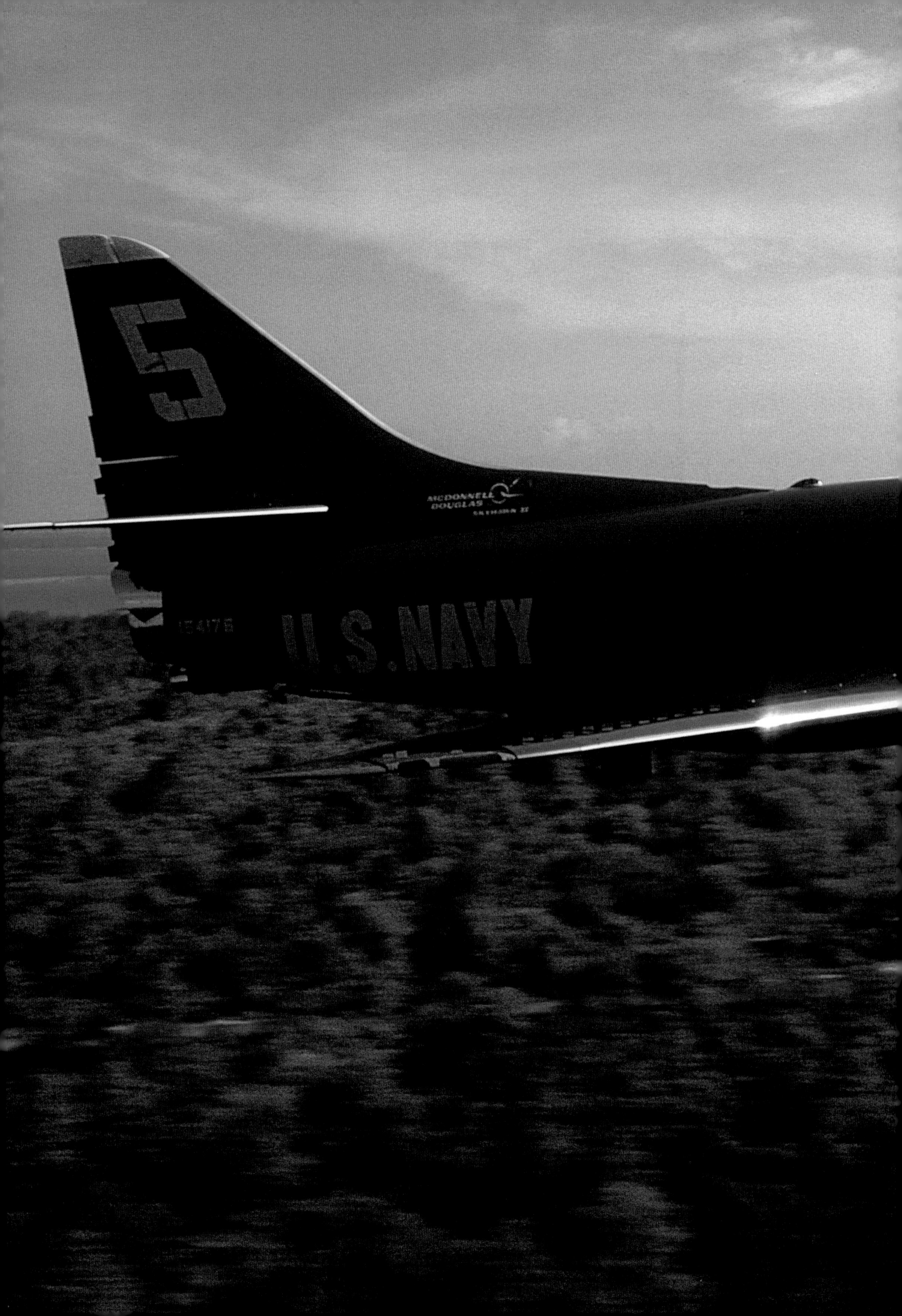
5
McDONNELL
DOUGLAS
U.S.NAVY

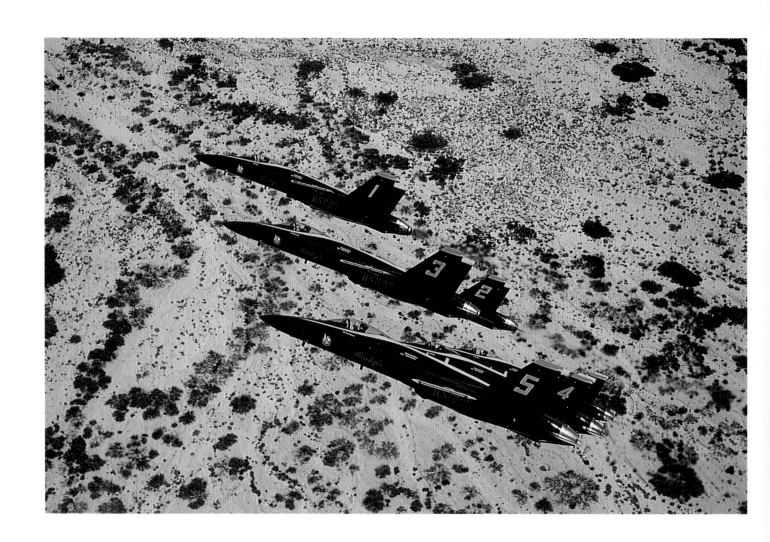

Super Blue

Caption/Data

■

Special Thanks

Flygstaben Informationsavd/Flygvapnet/スウェーデン空軍司令部広報課
Information Section/FDFGH/フィンランド国防軍総司令部広報課
Information Section/KdoFuF/スイス国防軍対航空機軍団広報課
Media Liaison Office/USFJ/在日アメリカ軍司令部報道連絡事務所
Naval Information Office/USN/アメリカ海軍広報室
Public Affairs Division/ANGB/アメリカ州空軍事務局広報部
Public Affairs Division/USAF/アメリカ空軍広報部
Public Affairs Office/USNR/アメリカ海軍予備役航空隊広報室
Public Information/Northrop Aircraft Group/ノースロップ航空機広報課
Public Information Office/JASDF/防衛庁航空幕僚本部広報室
Public Relation/Aermacchi S.p.A./アエルマッキ株式会社広報課
Public Relations/Klu/オランダ空軍広報部
Public Relations/RAFSC/イギリス空軍支援航空軍団広報部
Service d'Information et de Relation Publique des Armees/AdlA/フランス空軍司令部広報課
Ufficio Documentation e Attirita Promozionali/AMI/イタリア空軍司令部広報課
Blue Angels/USN/アメリカ海軍ブルーエンジェルズ
Blue Impulse/JASDF/航空自衛隊戦技研究班
Escadrilha da Fumaca/FAB/ブラジル空軍エスカドリラ・ダ・フマサ
Frecce Tricolori/AMI/イタリア空軍フレッチェ・トリコローリ
Patrouille de France/AdlA/フランス空軍パトルイユ・ド・フランス
Red Arrows/RAF/イギリス空軍レッドアローズ
Snowbirds/CAF/カナダ国防軍スノーバーズ
Thunderbirds/USAF/アメリカ空軍サンダーバーズ
Akiko Ozawa/DACT/小沢明子

■

Staff

Photographs & Texts/ 徳永克彦 Katsuhiko Tokunaga
Editorial Direction/佐野総一郎 Sohichiroh Sano
Art Direction/望月秀峻 Hidetaka Mochizuki
Design/伊藤直子 Naoko Ito + HOT ART

McDonnell Douglas F-15A Eagle

In a late Alaskan sunrise, a big F-15A Eagle looms in silence. Its smooth form makes the F-15 a difficult photographic subject, but when shown in silhouette, as in this picture, it takes on a surprisingly angular shape.

Nikon F2, 25-50mm Zoom, 1/30sec, F5.6, KM
Feb. 83/Elmendorf AFB, Alaska, U.S.A.

McDonnell Douglas F-15A Eagle

When the temperature falls below -40°C, it becomes ''painful,'' instead of simply ''cold.'' But the ground crew, or the ''Eagle Keepers,'' as they are called, begin preparing for the first flight long before sunrise.

Nikon F3, 300mm, 1/250sec, F5.6, KR
Feb. 83/Elmendorf AFB, Alaska, U.S.A.

McDonnell Douglas F-15A Eagle

Under the ground crew's guidance, an F-15A Eagle heads toward the arming area. Its huge intake seems to suggest the F-15's enormous power. The snow on the ground portrays the quietness of the area.

Nikon F3T, 400mm, 1/500sec, F5.6+2/3, KR
Feb. 83/Elmendorf AFB, Alaska, U.S.A.

McDonnell Douglas F-15A Eagle

When the control tower gives clearance to the F-15A Eagles, they begin heading for the runway. They seem awkward on the ground, but once they gain power and fly through the blue sky, they truly transform into ''Eagles.''

Nikon F2, 25-50mm Zoom, 1/250sec, F5.6, KM
Feb. 83/Elmendorf AFB, Alaska, U.S.A.

McDonnell Douglas F-15A Eagle

During the arming operation, the F-15A Eagle's long shadows stretch over the snow on the runway. The fact that the hours of daylight in Alaska during the winter are short makes work hard for both pilots and photographers.

Nikon F2, 25-50 Zoom, 1/125sec, F5.6, KM
Feb. 83/Elmendorf AFB, Alaska, U.S.A.

Raytheon/Ford AIM-9L Sidewinder

Even an air-to-air missile placed under the F-15's wing freezes in the cold weather of Alaska. The AIM-9L Sidewinder can also be fired in a head-on situation, using infrared rays or not.

Nikon F3, 300mm, 1/250sec, F5.6, KR
Feb. 83/Elmendorf AFB, Alaska, U.S.A.

McDonnell Douglas F-15A Eagle

F-15 Eagles are loaded with two F-100 engines, which enable them to develop over 20 tons of thrust. When the flame of the afterburners licks across the frozen runway, snow and ice turn to vapor in an instant.

Nikon F3, 180mm, 1/250sec, F2.8+1/2, KR
Feb. 83/Elmendorf AFB, Alaska, U.S.A.

McDonnell Douglas F-15A Eagle

Four F-15A Eagles fly through an Alaskan mountain range. The F-15 is constructed with all the modern technology available, yet its mechanical beauty dims before Alaska's majestic natural beauty.

Nikon F2, 50mm, 1/250sec, F8, KM
April 83/over Alaska Range, Alaska, U.S.A.

Grumman F-14A Tomcat

The F-14A Tomcat gained great notoriety in the movie, *Top Gun*. The aircraft's most notable characteristic is its unique variable-sweep wing, which enhances the Tomcat's ability to fly at supersonic speeds. When its wings are placed back, as shown in this photograph, the Tomcat looks almost like a delta-wing fighting jet.

Nikon F3T, 400mm, 1/500sec, F8, PKR
Nov. 85/NAS Pensacola, Florida, U.S.A.

F/A-18A Hornet & F-14A Tomcat

Over the Pacific Ocean just off the coast of California, two F-14A Tomcats and an F/A-18A Hornet do a splendid turn. Their simple gray color, for low visibility, sharply contrasts with the variegated camouflage of earlier Navy fighters.

Nikon F3P, 50mm, 1/250sec, F5.6+1/2, PKM
Feb. 87/over Channel Island, California, U.S.A.

Grumman F-14A Tomcat

An F-14A Tomcat leaves Miramar Naval Air Station in California. The bright flame bursting from its afterburners suggests the huge Tomcat engines.

Nikon F3T, 400mm, 1/500sec, F5.6+2/3, PKR
Feb. 86/NAS Miramar, California, U.S.A.

Grumman F-14A Tomcat

Traditionally, American jets are larger in size and more comfortable than those of the Europeans. The F-14A Tomcat, however, is extremely large in size. In this picture, the aircraft seems to dwarf a small female groundcrew member.

Nikon F3P, 180mm, 1/250sec, F5.6+1/2, PKM
Feb. 87/NAS Point Mugu, California, U.S.A.

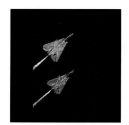

Grumman F-14A Tomcat

Over the Pacific Ocean, two F-14A Tomcats leave long JP-5 trails. While low-visibility gray is their paint scheme, the picture shows how ground crews moving about on the surface of the airplanes left a mottled design.

Nikon F3P, 50mm, 1/250sec, F5.6, PKM
Feb. 87/over Channel Island, California, U.S.A.

Naval Air Station Oceana

While the major U.S. naval aviation base for the Pacific area is Miramar, Naval Air Station Oceana, in Virginia, is the major base for the Atlantic Ocean area. On the snow-covered flight line, more than 100 jet fighters are visible.

Nikon F2, 200mm, 1/250sec, F8, KM
Feb. 80/NAS Oceana, Virginia, U.S.A.

McDonnell Douglas F-101B Voodoo

For aerial pictures of fighter jets, showing the most difficult maneuvers is a major goal. For the F-101B Voodoo, which achieved some of the fastest supersonic speeds ever recorded, turns and rolls are next to impossible. Here, a Voodoo attempts to pull a tight turn.

Nikon F2, 28mm, 1/250sec, F5.6+1/2, KM
Feb. 81/over Western New York State, U.S.A.

General Dynamics F-106A Delta Dart

The F-106A Delta Dart, the last American jet fighter that was a pure interceptor, brakes with its drag chute. A golden Massachusetts sunset burnishes the Delta Dart's fuselage.

Nikon F3T, 400mm, 1/250sec, F5.6+2/3, KR
Oct. 81/Otis ANGB, Massachusetts, U.S.A.

General Dynamics F-106A Delta Dart

Over the East Coast resort of Cape Cod fly four F-106 Delta Darts of the Massachusetts Air Guard. The pale fall sunlight emphasizes the sharp profiles of the airplanes.

Nikon F2, 28mm, 1/250sec, F5.6+2/3, KM
Oct. 81/over Cape Cod, Massachusetts, U.S.A.

Fairchild F-105D Thunderchief

The F-105D Thunderchief served as America's main fighter jet in the Vietnam war, but they are now retired from active service. This Thunderchief has just finished its mid-air refueling training and escorts a KC-135 back home to its Virginia air base.

Nikon F2, 50mm, 1/60sec, F4, KR
Jan. 80/over North Carolina, U.S.A.

McDonnell Douglas F-4D Phantom II

An early night mission for the F-4D Phantoms. As the sun goes down, they take off from McConnell Air Force Base. Captured here is the silhouette of an F-4D Phantom II following the preceding two-man formation.

Nikon F3P, 25-50mm Zoom, 1/125sec, F5.6, PKM
May 87/over Wichita, Kansas, U.S.A.

F-4C Phantom II and B-52G Stratofortress

An F-4C Phantom II practices aerial engagement with SAC's B-52G Stratofortress, which simulates an enemy aircraft. The F-4 assumes the ''ID'' position by sliding under the bomber. The ID position is used against Soviet bombers to identify their fuselage serial number.

Nikon F3, 25-50mm Zoom, 1/125sec, F5.6+1/2, KM
March 84/over Lake Huron, Michigan, U.S.A.

McDonnell Douglas F-4S Phantom II

An F-4S Phantom II rolls on its back in the sky over rough Texas terrain. The F-4 is painted in a tactical camouflage scheme developed by the U.S. Navy. It allows the fighter to minimize the reflection of sunlight even at the sharpest angles of attack, and is very effective.

Nikon F3P, 50mm, 1/250sec, F5.6, PKM
Feb. 86/over Grand Prairie, Texas, U.S.A.

McDonnell Douglas F-4S Phantom II

With the ground crew's guidance, an F-4S Phantom lands. The deep metallic blue of its fuselage, painted to celebrate the U.S. Navy's 75th year of aviation, contrasts beautifully with the white drag chute.

Nikon F3P, 180mm, 1/500sec, F4+1/2, PKM
Nov. 86/NAS Point Mugu, California, U.S.A.

McDonnell Douglas F-4C Phantom II

A wingman rolls 90 degrees as he follows his leader, breaking from a vertical climb. Two F-4C Phantoms of the Louisiana Air Guard descend, clearly showing their camouflage colors.

Nikon F3P, 50mm, 1/250sec, F5.6, PKM
Nov. 84/over Chandeleur Is., Louisiana, U.S.A.

McDonnell Douglas F-4E Phantom II

Since the Vietnam war, the F-4E Phantom II has been equipped with 20mm Balkan guns under its nose. For this reason, the F-4E Phantom IIs have a much more attractive-looking nose than the more unrefined F-4 Phantom I series.

Nikon F3T, 400mm, 1/250sec, F5.6+2/3, PKR
Feb. 86/Tyndall AFB, Florida, U.S.A.

McDonnell Douglas F-4D Phantom II

Looking forward to his flight, a Kansas Air Guard crew climbs aboard an F-4D Phantom II. To emphasize the black shine of its large nose, I aimed straight on with a 400mm super-telephoto lens.

Nikon F3T, 400mm, 1/500sec, F5.6+2/3, PKR
May 87/McConnell AFB, Kansas, U.S.A.

Fairchild A-10A Thunderbolt II

Two A-10A Thunderbolt IIs, or ''tankbusters,'' fly behind their flight leader. All of their armament is devoted to their tank-destroying mission, so they should appear beautiful when performing their function, but....

Nikon F2, 85mm, 1/250sec, F5.6, KM
July 82/over Gettysburg, Pennsylvania, U.S.A.

Fairchild A-10A Thunderbolt II

While the morning mist still rises at England Air Force Base, an A-10A Thunderbolt II receives final arming preparations before its takeoff. The huge engine pods, the double vertical stabilizers, and the aircraft's peculiar nose, seem to overwhelm the ground crew.

Nikon F3T, 400mm, 1/250sec, F5.6+2/3, KR
March 82/England AFB, Louisiana, U.S.A.

McDonnell Douglas F/A-18A Hornet

An F/A-18A's glass cockpit reflects the instrument panel in the dark. This modern instrumentation, which includes DDI made up of large-sized CRTs and HUD, will be the standard cockpit for future jet fighters.

Pentax 645, 45mm, 8sec, F5.6, PKR
Feb. 87/NAF El Centro, California, U.S.A.

General Dynamics FB-111A

An FB-111A bomber breaks wide to the right after receiving its mid-air refueling. Its nickname is the ''aardvark,'' because of its long, curved nose, but its white belly reminds us of a reptile.

Nikon F2, 85mm, 1/250sec, F5.6+1/2, KM
June 80/over New Hampshire, U.S.A.

McDonnell Douglas F/A-18A Hornet

The cockpits of many jet fighters seem messy, because they are crammed with high-tech equipment. But the modern cockpit of the F-18 makes this impression a thing of the past.

Pentax 645, 45mm, 1/125sec, F5.6+1/2, PKR
Feb. 87/NAS Point Mugu, California, U.S.A.

Northrop F-20A Tigershark

Due to the shape of its nose, the F-20A of Northrop Company came to be known as the Tigershark. In order to emphasize its distinctive feature, I used a wide-angle 20mm lens and shot from overhead.

Nikon F3P, 20mm, 1/125sec, F8+1/3, PKM
Nov. 85/Edwards AFB, California, U.S.A.

Northrop F-20A Tigershark

An F-20A Tigershark flies over the vast Mojave Desert of California. The Tigershark's special color is called ''BMW gray,'' and is actually used on BMW automobiles.

Nikon F3P, 50mm, 1/250sec, F5.6+1/3, PKM
Sept. 84/over Mojave Desert, Calif., U.S.A.

Northrop F-20A Tigershark

An F-20A Tigershark in a vertical climb. Originally F-20As were designed to be sold to other countries. I thought that a Tigershark History Book might be published successfully, and this picture was to be on the front cover, but now...

Nikon F3P, 50mm, 1/250sec, F5.6 + 1/2, PKM
Sept. 84/over Mojave Desert, Calif., U.S.A.

Dassault-Brequet Mirage 2000C

At an altitude of 39,000 feet, a Mirage 2000C recovers from only 70 knots of airspeed. White clouds, "Super Blue" sky, and the delta wings of the Mirage 2000C make a splendid contrast.

Nikon F3P, 50mm, 1/250sec, F5.6 + 2/3, PKM
July 87/over Cote D'Or, Bourgogne, France

Dassault-Brequet Mirage 2000C

A French Air Force pilot awaits a signal from the cockpit of his Mirage 2000C. The serious expression on his face probably indicates that he is thinking about the day's mission.

Nikon F3P, 180mm, 1/125sec, F4, PKM
July 87/BA 102, Dijon, France

Dassault-Brequet Mirage 2000C

Two Mirage 2000Cs turn in formation. The Mirage possesses a trim, aerodynamic design, which is common to most French fighter jets. The refueling probe seems to be an appendage to the sleek Mirage.

Nikon F3P, 50mm, 1/250sec, F5.6 + 2/3, PKM
July 87/over Cote D'Or, Bourgogne, France

Dassault-Brequet Mirage 2000C

A Mirage 2000C and its pilot. Photographed in the normal way, the huge size of the Mirage overpowers the pilot, so in this picture, I showed only a portion of the Mirage in its hangar, with the sun shining on its nose.

Nikon F3P, 180mm, 1/125sec, F5.6 + 1/3, PKM
July 87/BA 102, Dijon, France

Saab-Scania AJ37 Viggen

An AJ37 Viggen demonstrates its superb short-distance landing capability as it employs its thrust-reverser upon touching down. Its drooping nose seems to suggest the power of the sudden braking when the reverse thrust is used.

Nikon F3T, 300mm, 1/500sec, F5.6 + 2/3, KR
Sept. 84/F14, Halmstad, Sweden

Saab-Scania JA37 Viggen

The JA37 Viggen's light gray nose is side lit. The aircraft seems to have a blunt nose, but when viewed up close, it looks surprisingly sharp and sleek.

Nikon F2, 200mm, 1/250sec, F5.6 + 1/3, KM
Aug. 83/F13, Norrköping, Sweden

Saab-Scania AJ37 Viggen

By having both canards and delta wings, the AJ37 Viggen is called the "Co-Delta" because of its unique design. Two AJ37s loop in formation over the fertile land of southern Sweden.

Nikon F2, 50mm, 1/250sec, F5.6 + 1/3, KM
Sept. 84/over Vastergotland, Sweden

Saab-Scania JA37 Viggen

A JA37 Viggen approaches F13 Air Base near Norrköping. While this Swedish Air Force base lies in a pastoral environment, the main facilities are all underground, ready for actual combat.

Nikon F2, 200mm, 1/250sec, F5.6 + 1/3, KM
Aug. 83/F13, Norrköping, Sweden

Saab-Scania AJ37 Viggen

With the roaring sound of its afterburners, an AJ37 Viggen takes off for a training flight. The light on the aircraft shows the curving, complex lines of its fuselage.

Nikon F3T, 300mm, 1/500sec, F5.6 + 2/3, KR
Sept. 84/F7, Sátenás, Sweden

Dassault-Brequet Mirage 4000

In this head-on photograph, a Mirage 4000 approaches the runway. The Mirage 4000 is a scaled-up model of the Mirage 2000; the body lines are much more feminine-looking because of its fat radome, containing extensive radar equipment.

Nikon F3T, 400mm, 1/500sec, F5.6 + 1/2, PKR
Sept. 86/RAE Farnborough, Hampshire, U.K.

McDonnell Douglas F-15C Eagle

An F-15C Eagle starts up its engines in front of a hangar. Even though its body and paint are identical to other U.S. Air Force jets, the small flowers in the foreground of the picture give it a distinctly European flavor.

Nikon F3T, 300mm, 1/500sec, F5.6 + 2/3, KR
May 82/Bitburg AB, Osbur Hochwald, W. Germany

Saab-Scania J35F Draken

Two double-delta-winged J35F Drakens take off in formation. The huge white wing numbers do not balance with the orthodox camouflage of the aircraft.

Nikon F3T, 400mm, 1/500sec, F5.6 + 2/3, KR
Sept. 84/F10 Angelhorm, Sweden

Saab-Scania SK35C Draken

The Rovaniemi Air Base, located in Lapland, is the most northern air force base in Finland. Even though the picture was taken in summer, it appears to be cold because of the proximity of the Arctic Circle.

Nikon F2, 25-50mm Zoom, 1/60sec, F5.6, KM
May 83/Rovaniemi, Lapland, Finland

Saab-Scania SK35C Draken

An SK35C Draken of the Finnish Air Force taxis on the tarmac between stands of conifers. The airplane itself is a dull gray color, but the bright sun and clear air of the region give the picture a feeling of tension and readiness.

Nikon F3T, 300mm, 1/500sec, F5.6 + 2/3, KR
May 83/Rovaniemi, Lapland, Finland

J35F/SK35C Draken & SK60

With Sweden's brilliant sky as a backdrop, J/SK35s and SK60s break away. It is the 40th anniversary of the F16 base. Big formations like this are becoming rarer as time passes by.

Nikon F2, 25-50mm Zoom, 1/250sec, F5.6, KM
Aug. 83/F16, Uppsala, Sweden

General Dynamics F-16B Fighting Falcon

The weather of the Netherlands is always unsuitable for sharp photographs, so in this situation, where there was still some hazy light left, I went more for the landscape instead of the jet itself. Leeuwarden, where this picture was taken, usually has bad weather because it is near the coast.

Nikon F2, 200mm, 1/125sec, F5.6, KM
May 82/Leeuwarden, The Netherlands

General Dynamics F-16A Fighting Falcon

The maneuverability of the F-16A Fighting Falcon is incredibly good. The smoke trailing from the wings and the angle of the jet's position enabled me to show its amazing maneuverability in a still picture.

Nikon F3T, 400mm, 1/500sec, F5.6 + 1/2, PKR
June 85/Ramstein AB, Pfälzer Wald, W. Germany

General Dynamics F-16 Fighting Falcon

At the Paris-Salon, the biggest airplane show in the world, an F-16C Fighting Falcon begins its demonstration flight. In these airshows, where jet fighters exhibit their maximum capabilities, photographers get numerous valuable photo opportunities.

Nikon F3T, 400mm, 1/500sec, F8, PKR
June 85/AP Le Bourget, Paris, France

McDonnell Douglas Phantom FGR.2

A Phantom FGR.2 returns to Coningsby Royal Air Force Base in Lincolnshire. The flowery scene is reminiscent of a peaceful, tranquil fantasy-land, but Coningsby is the central base in England and one of the most important RAF bases in the world.

Nikon F3T, 400mm, 1/500sec, F5.6 + 2/3, KR
June 81/RAF Coningsby, Lincolnshire, U.K.

Fairchild A-10A Thunderbolt II

An A-10A Thunderbolt II flies through the Swiss Alps. Its camouflage, selected for the bad weather in Europe, enables this American-made attack jet to blend with the scenery quite well.

Nikon F3T, 400mm, 1/250sec, F5.6 + 2/3, KR
June 82/FP Sion, Switzerland

Aermacchi MB. 339B

An MB. 339B dives inverted toward Lake Como in Italy. The purpose of this photograph was to get the reflection from the lake, but the camera focus was the land instead. The beautiful coastline helped make this picture come alive.

Nikon F3P, 85mm, 1/250sec, F5.6 + 1/2, PKM
May 85/over Lake Como, Italy

Aermacchi MB. 339A

An MB. 339A takes off for a demonstration flight in the airshow in Sion, Switzerland. In order to give a fresh look to the picture, I altered the height from which the photo was taken. In this case, I raised the viewpoint a little by climbing to the top of a hangar.

Nikon F3T, 300mm, 1/500sec, F5.6 + 2/3, KR
June 82/FP Sion, Switzerland

British Aerospace Harrier GR.Mk.3

Looking down on the mountains, a Harrier GR.Mk.3 takes off. Its camouflage, intended to allow the fighter to blend with the background, is a big problem for a photographer. However, the brief glint on the fuselage made a beautiful highlight.

Nikon F3T, 400mm, 1/500sec, F5.6 + 2/3, PKR
June 86/FP Sion, Switzerland

Lockheed SR-71A

In the present, even after 200 years of manned flight, the SR-71A, a Mach-3 strategic reconnaissance aircraft, remains mysterious and exotic. On takeoff it has incredible power, shooting flames from the afterburners that are beyond compare to any other jet.

Nikon F3T, 400mm, 1/500sec, F5.6 + 1/2, PKR
Sept. 86/RAE Farnborough, Hampshire, U.K.

Mikoyan MiG-21bis Fishbed N

Soviet troops always stand in the way when I photograph their military jets. Getting permission to photograph Soviet airplanes is extremely difficult. This picture of a MiG-21 was finally taken during my third visit to Finland.

Nikon F2, 25-50mm Zoom, 1/125sec, F8 + 1/3, KM
Aug. 84/Vantaa AP, Helsinki, Finland

Fouga Magister

A Magister inside the hangar reminds us of an insect hiding in a cellar. This French training plane does not even have an ejection seat; however, it has a great reputation for its reliable design, and has become a best-seller throughout Europe.

Nikon F3T, 300mm, 1/125sec, F5.6 + 1/2, KR
May 83/Rovaniemi, Lapland, Finland

Dassault-Brequet Alpha Jet E

With the Alps at its back, an Alpha Jet negotiates a turn. Normally photographers avoid taking pictures into the sun, but in this situation, with the silver-colored jet in shadow, I was able to capture the effect of Japanese sumie-style painting.

Nikon F3T, 400mm, 1/500sec, F5.6 + 2/3, PKR
June 86/FP Sion, Switzerland

Panavia Tornado IDS

When I requested permission to photograph the Tornado, the mountain ranges of Northern Italy were a must for the background. In reality, there were not as many photo opportunities as I had expected, but this kind of landscape can be found only in Europe.

Nikon F3P, 50mm, 1/250sec, F5.6 + 1/2, PKM
April 85/over Italian Alps, Italy

Tornado & F-104S Starfighter

An F-104S Starfighter escorts a Tornado. The F-104 has been described as the last ''human'' fighter aircraft. Most NATO countries are now retiring the fighter. Currently the F-104S is the only model that may be found in Europe, and it is flown only by the Italian Air Force.

Nikon F3P, 50mm, 1/250sec, F5.6 + 1/2, PKM
April 85/over Tarrant Bay, Italy

Panavia Tornado IDS

A Tornado IDS flies above cloud cover. Since Italy prohibits photographing the ground from the air, most aerial pictures are made over clouds.

Nikon F3P, 50mm, 1/250sec, F5.6 + 1/2, PKM
April 85/over Tarrant Bay, Italy

McDonnell Douglas F-15C Eagle

After mid-air refueling, an F-15C Eagle breaks to the left, yielding its position to other fighters. Below is the shining surface of the sea at sunrise, with shadows of cloud formations. These contrasts can be seen most clearly from 25,000 feet above the ocean.

Nikon F2, 28mm, 1/250sec, F5.6 + 1/3, KM
April 80/over East China Sea, Okinawa, Japan

McDonnell Douglas F-15C Eagle

A tanker, which furnishes ideal angles from the front and even from the top, is one of the best camera platforms. The large form of this F-15 Eagle fills the KC-135's window for mid-air refueling.

Nikon F3P, 25-50mm Zoom, 1/125sec, F5.6+1/2, KM
Oct. 82/over East China Sea, Okinawa, Japan

McDonnell Douglas F-15J Eagle

In cold Hokkaido, Japan, an F-15J on approach is captured from behind in a metal-colored sky. The wings and vertical stabilizers are straight, with no curves. Surprisingly, the F-15, one of the most effective modern-day fighters, is very simple in construction.

Nikon F3T, 400mm, 1/250sec, F5.6+2/3, KR
Jan. 84/Chitose AB, Hokkaido, Japan

McDonnell Douglas F-15DJ Eagle

An F-15DJ Eagle of the Air Self Defense Force practices the touch-and-go maneuver. The photograph is taken from a common vantage point, but the huge speed brakes on the F-15 make the picture striking.

Nikon F3T, 400mm, 1/500sec, F5.6+2/3, PKR
May 86/Nyutabaru AB, Miyazaki, Japan

McDonnell Douglas F-15C Eagle

This photograph caught an F-15C Eagle flying through the sky during a canopy roll. There is only one chance for the shutter in aerophotographing a rolling jet, but in order to achieve different angles, it is a valuable technique.

Nikon F3P, 50mm, 1/250sec, F5.6+1/2, PKM
April 85/over East China Sea, Okinawa, Japan

McDonnell Douglas F-15D Eagle

Air-to-air gunnery. An F-15D Eagle follows the dart target pulled by an F-86 Sabre. Positioned in the dark blue sky, the target shines silver. Targets are clearly visible through the huge bubble canopy.

Nikon F3P, 16mm, 1/250sec, F5.6+1/2, PKM
April 85/over East China Sea, Okinawa, Japan

McDonnell Douglas F-15C Eagle

A U.S. Air Force F-15C Eagle taxis at Kadena Air Base, Japan. The lowered intake lip is part of the F-15's special system to control the amount of air that enters the engines.

Nikon F3T, 300mm, 1/500sec, F5.6+2/3, PKR
Oct. 82/Kadena AB, Okinawa, Japan

McDonnell Douglas F-15DJ Eagle

This fighter has just returned to its base after a long night flight. The most important element in night photography is the contrast of the dark sky and the shining fighter. Both the aircraft and the guiding lights should be used positively.

Nikon F3T, 85mm, 40sec, F5.6, PKR
May 86/Nyutabaru AB, Miyazaki, Japan

Rockwell RA-5C Vigilante

On board the U.S.S. *Kitty Hawk*, just off the coast of the Philippines, an RA-5C Vigilante is prepared for a catapult launch. The number of deck crewmen working in the limited space is surprisingly large.

Nikon F2, 50mm, 1/250sec, F5.6+1/2, KM
Feb. 78/U.S.S. *Kitty Hawk*, in East China Sea, Ph.

Mitsubishi F-1

Misawa Air Base has one of the highest snow accumulations of any air base in Japan; however, the schedule for practice flights does not change. Just as soon as the snow stops falling, the F-1's engines are started up.

Nikon F3P, 180mm, 1/250sec, F5.6+1/3, PKM
Jan. 85/Misawa AB, Aomori, Japan

General Dynamics F-16A/B Fighting Falcon

A group of F-16A/B Fighting Falcons line up noses in a row. Ever since the F-4D pulled off the base, Misawa Air Base has been called "sleepy hollow," but the arrival of F-16s in 1985 gave new life to the base.

Nikon F3T, 300mm, 1/500sec, F5.6+2/3, PKR
June 86/Misawa AB, Aomori, Japan

General Dynamics F-16A Fighting Falcon

The tremendous power contained in modern fighter jets such as this F-16A Fighting Falcon is the reason that the G-load does not diminish at the pinnacle of a loop. Wing vapors are evidence of that power.

Nikon F3P, 50mm, 1/250sec, F5.6+1/2, PKM
June 86/over Japanese Sea, Aomori, Japan

General Dynamics F-16A/B Fighting Falcon

In the cockpit of the F-16A Fighting Falcon, the pilot's waist is practically level with the edge of the fuselage. This, of course, offers an ideal view for the photographer.

Nikon F3P, 16mm, 1/250sec, F5.6+1/2, PKM
June 86/over Japanese Sea, Aomori, Japan

Lockheed F-104J Starfighter

The Starfighter's cockpit is relatively small. While it is adequate for the small-sized Japanese pilots, the NATO air force pilots must have difficulties trying to fit inside.

Nikon F3T, 300mm, 1/500sec, F5.6+1/2, PKR
June 84/Naha AB, Okinawa, Japan

Fuji T-1A

The T-1A trainer was the first jet to be manufactured in post-war Japan. The influence of the F-86 Sabre, a popular aircraft at the time, can be seen in design elements, but in overall form, it is a true Japanese trainer.

Nikon F3P, 50mm, 1/250sec, F5.6+1/2, PKM
July 86/over Genkainada, Fukuoka, Japan

Fuji T-1A

A Fuji T-1A flies over Genkainada, near its home base, Ashiya Air Base. The clouds reflecting off the sea surface give an interesting texture to the photograph.

Nikon F3P, 50mm, 1/250sec, F5.6+2/3, PKM
July 86/over Genkainada, Fukuoka, Japan

General Dynamics F-16A Fighting Falcon

Off the coast of Misawa over the Japan Sea, U.S. Air Force F-16 Fighting Falcons pull a tight turn. This northern sea is the same two gray colors as the jets. They both suggest the freezing-cold water.

Nikon F3P, 50mm, 1/250sec, F5.6+1/2, PKM
June 86/over Japanese Sea, Aomori, Japan

McDonnell Douglas F-4D Phantom II

Loaded with Maverick missiles, an F-4D Phantom II taxis out for a training flight. An approaching tropical storm in the background accents the bright light on the aircraft.

Nikon F2, 400mm, 1/250sec, F5.6+1/2, KR
Aug. 79/Kadena AB, Okinawa, Japan

McDonnell Douglas F-4S Phantom II

Extreme care is required to photograph the upper part of a subject aircraft. Since the lift from the wings of the subject aircraft and the photographing aircraft cross each other, an air collision could easily result from the slightest error.

Nikon F3P, 50mm, 1/250sec, F5.6+1/3, PKM
July 85/over Pacific Ocean, Japan

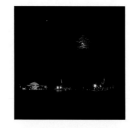

McDonnell Douglas F-4E/J Phantom II

F-4E/Js are lined up on the flight line for a night mission at Nyutabaru Air Base, Miyazaki, Japan. Moonlight is usually bright and effective for picture-taking, but it is difficult to position the elements. The cloud helped the composition of this picture.

Nikon F3P, 25-50mm Zoom, 4sec, F4, PKR
May 86/Nyutabaru AB, Miyazaki, Japan

McDonnell Douglas F-4D Phantom II

Slow shutter speed is useful when you want to emphasize aircraft speed. It can also be used to add an accent to an image, as in this picture of a crew about to board their beloved F-4D Phantom II at Kadena Air Base, Okinawa, Japan, around sunset.

Nikon F2, 135mm, 1/15sec, F5.6, KR
Aug. 79/Kadena AB, Okinawa, Japan

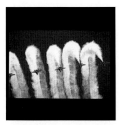

McDonnell Douglas F/A-18A Hornet

The Blue Angels demonstrate a "line abreast loop" with five aircraft in a row. Smoke trails make the demonstration more impressive. This is a difficult maneuver because there is no reference point between the individual aircraft.

Nikon F3T, 400mm, 1/500sec, F5.6+2/3, PKR
May 87/NAS Cecil Field, Florida, U.S.A.

McDonnell Douglas F/A-18A Hornet

"Diamond 360." The Blue Angels switched to F/A-18 Hornets in 1987. They still maintain their traditional super-tight formations—the distance between the aircraft is under 90cm.

Nikon F3T, 400mm, 1/500sec, F5.6+2/3, PKR
March 86/NAF El Centro, California, U.S.A.

McDonnell Douglas F/A-18A Hornet

To the American aerobatic team, the show starts on the ground before the flight. It is quite the opposite for the professional European teams, who concentrate only on their flying.

Nikon F3P, 180mm, 1/250sec, F5.6+1/2, PKM
Feb. 86/NAF El Centro, California, U.S.A.

McDonnell Douglas A-4F Skyhawk II

They are both American teams, but the Navy Blue Angels are more aggressive than the Air Force Thunderbirds. This A-4's altitude is so low that it nearly touches the trees. The photograph was taken at a slow shutter speed.

Nikon F3P, 25-50mm Zoom, 1/15sec, F16+2/3, PKM
May 85/over Choctow, Florida, U.S.A.

McDonnell Douglas F/A-18A Hornet

The Blue Angels fly above the southern California desert near the Mexican border. They conduct intensive training here in the desert every year for three months before the show season.

Nikon F3P, 50mm, 1/250sec, F5.6+2/3, PKM
April 87/over Imperial Valley, Calif., U.S.A.

McDonnell Douglas A-4F Skyhawk II

Four A-4 Skyhawk IIs roll together as one body. The Blue Angels dislike having a passenger in their formation. Building up trust with them is the first step.

Nikon F3P, 16mm, 1/250sec, F5.6+1/3, PKM
May 85/over Choctow, Florida, U.S.A.

F/A-18A Hornet & T-34C Turbo Mentor

The Blue Angels took delivery of the F/A-18 Hornet in 1987. They now fly in a large performance envelope. In a still picture it is hard to show the low-speed capability compared with the high-speed capability. I tried to demonstrate it by having a low-speed T-34 fly alongside the Hornet.

Nikon F3P, 50mm, 1/250sec, F5.6+2/3, PKM
April 87/over Imperial Valley, Calif., U.S.A.

CT-114, Alpha Jet, MB. 339A, A-4F, & T-27

After getting permission from each country, we finally composed a mixed formation with five of the world's top teams. It probably was the most difficult flight to arrange in my entire career.

Nikon F3P, 50mm, 1/125sec, F4+1/3, PKM
Aug. 86/over Abbotsford, B.C., Canada

General Dynamics F-16A Fighting Falcon

Looking down on a Thunderbird F-16A Fighting Falcon from a mid-air refueling tanker. You can see that the pilot wears a brightly colored helmet, as colorful as the body of the aircraft.

Nikon F2, 85mm, 1/125sec, F5.6+2/3, KM
June 83/over Ft. Worth, Texas, U.S.A.

General Dynamics F-16A Fighting Falcon

The ground crewman dashes with chocks to the F-16 Fighting Falcon that just taxied in. While it is exciting for Americans viewing the show, it is a little too much for a Japanese audience.

Nikon F3P, 180mm, 1/250sec, F5.6+1/3, PKM
Feb. 85/Nellis AFB, Nevada, U.S.A.

General Dynamics F-16A Fighting Falcon

The Thunderbirds' delta formation demonstrates a 360-degree turn. The big blue birds that are painted on the bellies of the aircraft suggest the Americans' taste for ostentation.

Nikon F3T, 400mm, 1/500sec, F8, PKR
June 84/RAF Upper Heyford, Oxfordshire, U.K.

Northrop T-38A Talon

A T-38A Talon has just arrived at Indian Springs, where the Thunderbirds' training base is located. The polished body gleams in the early morning sunlight.

Nikon F3T, 400mm, 1/500sec, F8, PKR
May 83/Indian Springs AAF, Nevada, U.S.A.

Mitsubishi T-2
The ''Blue Impulse T-2'' starts with a delta roll displaying the five different smoke colors. It is rare to use five colors at the same time—the relatively minor contrast might be considered a weakness.
Nikon F3T, 400mm, 1/500sec, F5.6 + 2/3, PKR
Nov. 85/Yokota AB, Tokyo, Japan

British Aerospace Hawk T.Mk.1
England has a long history of aerobatic teams. In fact, the present team, the Red Arrows, demonstrates a high level of proficiency. This is a ''wine-glass roll.''
Nikon F3P, 25-50mm Zoom, 1/250sec, F5.6 + 2/3, PKM
April 86/over RAF Akrotiri, Limassol, Cyprus

Canadair CT-114 Tutor
The Snowbirds' CT-114 Tutors gather speed for a loop. A few seconds after this photograph was made, we experience Gs in the pull up. Thanks to the side-by-side seats of the CT-114, I was able to photograph from this angle.
Nikon F3P, 20mm, 1/125sec, F5.6, PKM
April 86/over Vancouver Island, B.C., Canada

British Aerospace Hawk T.Mk.1
A breathtaking opposing-pass performance. Almost every team that has a two-aircraft solo show performs this maneuver, but the Red Arrows are the best, based on the accuracy of their crossing altitudes and the small distance between the airplanes.
Nikon F3T, 400mm, 1/500sec, F5.6 + 2/3, PKR
July 86/Kameyoran AP, Jakarta, Indonesia

Canadair CT-114 Tutor
The Snowbirds fly over Vancouver Island, Canada, in a ''Concord'' formation. Every nation has its own beautiful scenery, but this magnificent landscape is unique to Canada.
Nikon F2, 28mm, 1/250sec, F5.6 + 1/2, KM
April 82/over Vancouver Island, B.C., Canada

British Aerospace Hawk T.Mk.1
The finale of the Red Arrows' airshow is the one and only ''parasol break.'' For the aerobatic teams, bold breaks and swift join-ups are the keys to dynamic performances.
Nikon F3P, 180mm, 1/250sec, F5.6 + 1/2, PKM
Sept. 86/Duxford, Cambridgeshire, U.K.

Canadair CT-114 Tutor
The Snowbirds perform their rolls over the center pylon of an air racecourse. Since the team trails so much smoke, this type of wide-angle shot is fun for a photographer.
Nikon F3P, 16mm, 1/250sec, F5.6 + 1/2, PKM
June 84/CFB Moose Jaw, Saskatchewan, Canada

British Aerospace Hawk T.Mk.1
A ''must'' for the aerophotography of aerobatic teams is this kind of vertical shot. Since there is a great amount of G-force present, it is difficult for a photographer, but when you succeed, the satisfaction you feel is unbelievable.
Nikon F3P, 50mm, 1/250sec, F5.6 + 1/2, PKM
April 86/over RAF Akrotiri, Limassol, Cyprus

Aermacchi MB. 339A/PAN
Ten MB. 339A/PANs, trailing smoke in the Italian national colors, come over the Swiss Alps. The Frecce Tricolori's thrilling and dynamic airshow has begun.
Nikon F3T, 400mm, 1/500sec, F5.6 + 1/2, PKR
June 86/Sion AP, Sion, Switzerland

Aermacchi MB. 339A/PAN
The Frecce Tricolori's MB. 339A/PANs become inverted at the top of a loop. The delta-shaped runway on the ground provides a sharp contrast to the Italian national colors of the aircraft.
Nikon F3P, 20mm, 1/250sec, F5.6 + 1/2, PKM
Aug. 86/over Abbotsford IAP, B.C., Canada

Aermacchi MB. 339A/PAN
One of the facinating aspects of aerobatic teams is that so many aircraft go up in the sky at once. Ten Frecce Tricolori MB. 339A/PANs approach a G.222 whose rear loading door is open.
Nikon F3P, 180mm, 1/500sec, F4, PKM
April 86/over Friuli, Italy

Aermacchi MB. 339A/PAN
Speaking of the Frecce Tricolori, their record-breaking, solo demonstrations are famous. This picture shows the ''Iomçovák,'' which means ''drunken man'' in Czechoslovakian. The pilot puts the aircraft into a flat inverted spin which is incredible to watch.
Nikon F3T, 400mm, 1/500sec, F5.6 + 2/3, PKR
Aug. 84/Klagenfurt AP, Klagenfurt, Austria

■**Abbreviation**／略語
AAF : Auxiliary Air Field／補助飛行場
AB : Air Base／航空基地
AFB : Air Force Base／空軍基地
ANGB : Air National Guard Base／州空軍基地
AP : Airport／**Aeroporto**／空港
BA : Base Aeriénne／航空基地
CFB : Canadian Forces Base／カナダ国防軍基地
F : Flygvapnet／スウェーデン空軍
IAP : International Airport／国際空港
NAF : Naval Air Facility／海軍航空施設
NAS : Naval Air Station／海軍航空基地
RAE : Royal Aircraft Establishment／イギリス航空機庁
RAF : Royal Air Force／イギリス空軍
USS : United States Ship／アメリカ艦艇

蒼空の視覚／*Super Blue*

■

1987年11月21日初版発行

著者／徳永克彦

発行人／鶴味政一

発行所／株式会社CBS・ソニー出版

住所／〒102 東京都千代田区五番町6番地2

電話／東京(03)234-5811

振替／東京1-65823

印刷・製本所／大日本印刷株式会社

定価／2800円

■

©1987 *KATSUHIKO TOKUNAGA*

Printed in Japan

乱丁、落丁本はお取替いたします